From the Author of Who Would Win?

JERRY PALLOTTA

WHAT WOULD YOU RATHER BE?

A TIGER OR A TIGER SHARK?

Thank you to Ocean Cousins:
Eddie, Paula, Sharon, and Sean.

Library of Congress Cataloging-in-Publication Data available
ISBN 978-1-339-03556-7
10 9 8 7 6 5 4 3 26 27 28 29
Printed in the U.S.A. 40
First edition, September 2025
Book design by Jaime Lucero

If you had to choose, would you rather be a tiger or a tiger shark?

A tiger is a mammal.

A tiger shark is a fish.

A tiger is a cat.
cat

A tiger shark is a shark.

A tiger lives on land.

A tiger shark lives in the ocean.

A tiger has four legs.

A tiger shark has eight fins.

A tiger's tail helps it turn when running fast.

A tiger shark's tail helps it swim.

A tiger has long teeth and a strong jaw.

A tiger shark has dozens of extra-sharp teeth.

A tiger has ears that stick up.
ears

A tiger shark has tiny holes for ears.

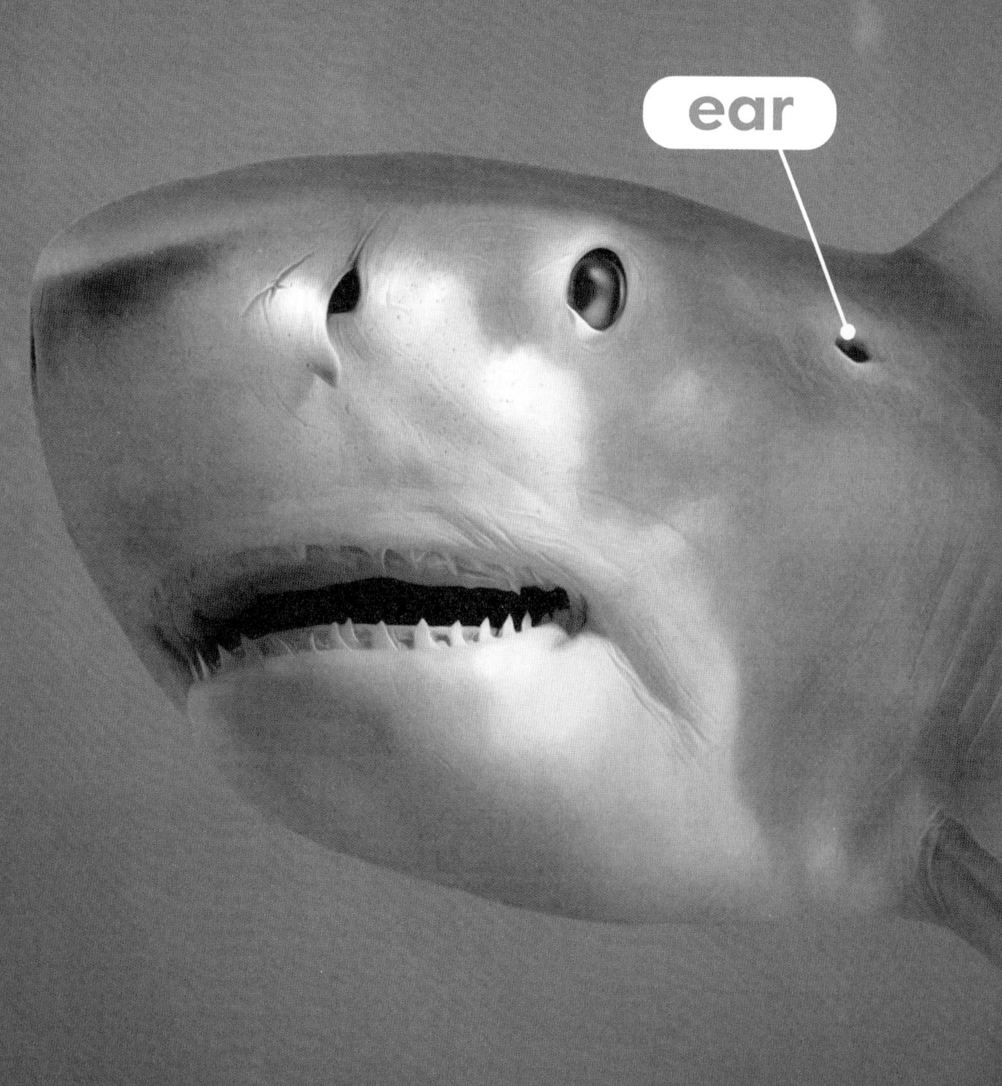

A tiger eats meat.

A tiger shark eats fish.

Tigers live alone.

Tiger sharks also live alone.

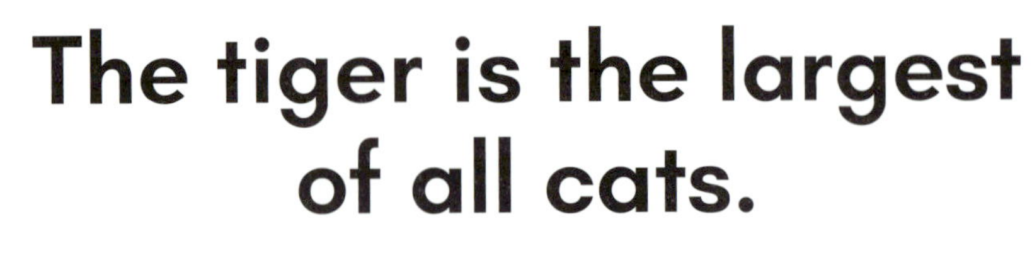

The tiger is the largest of all cats.

The tiger shark is the fourth-largest shark in the ocean.

A tiger has lungs.

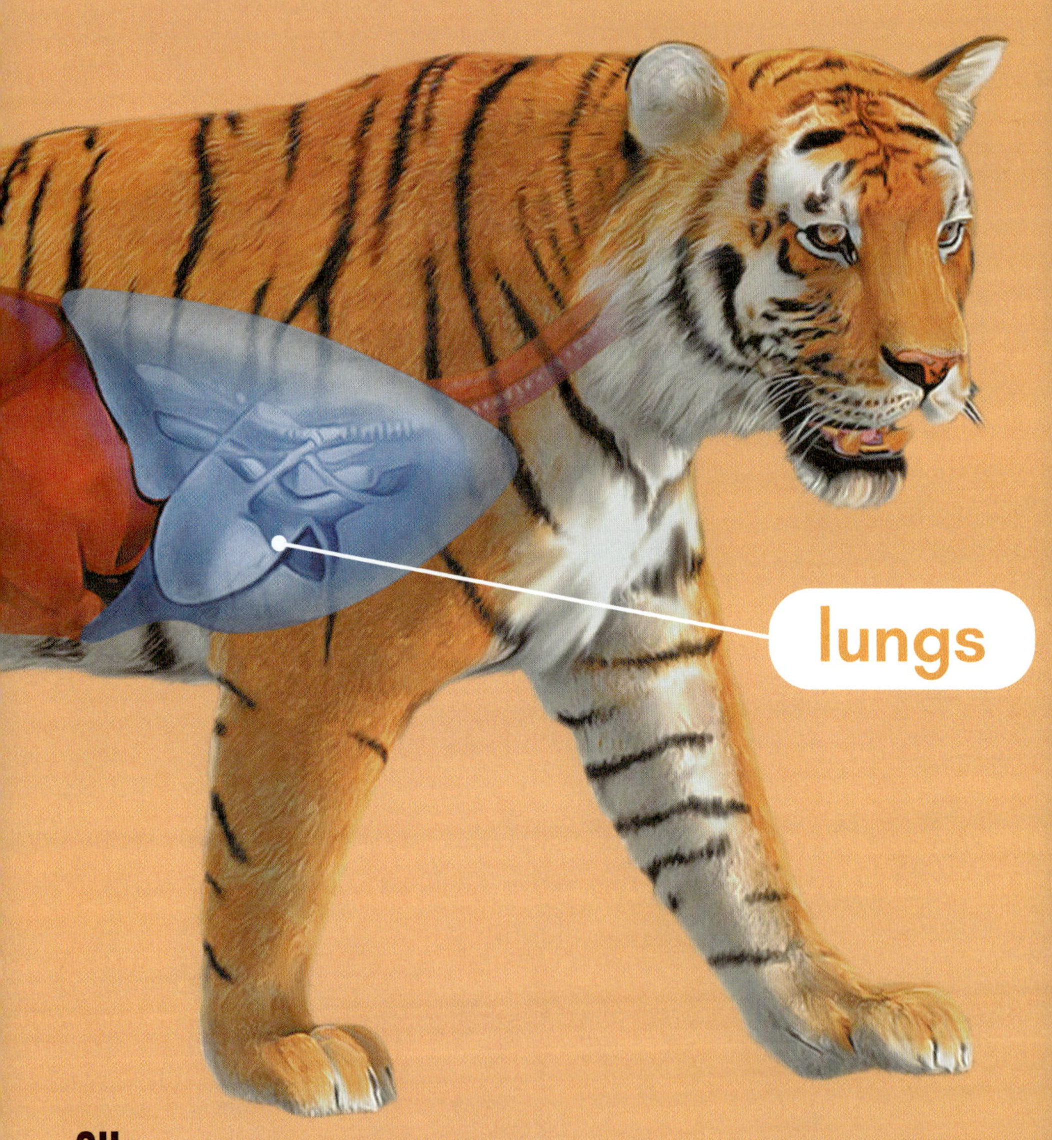

A tiger shark has gills.

A tiger has fur.

A tiger shark has rough skin.

A tiger has orange and black stripes.

A tiger shark has dark gray stripes.

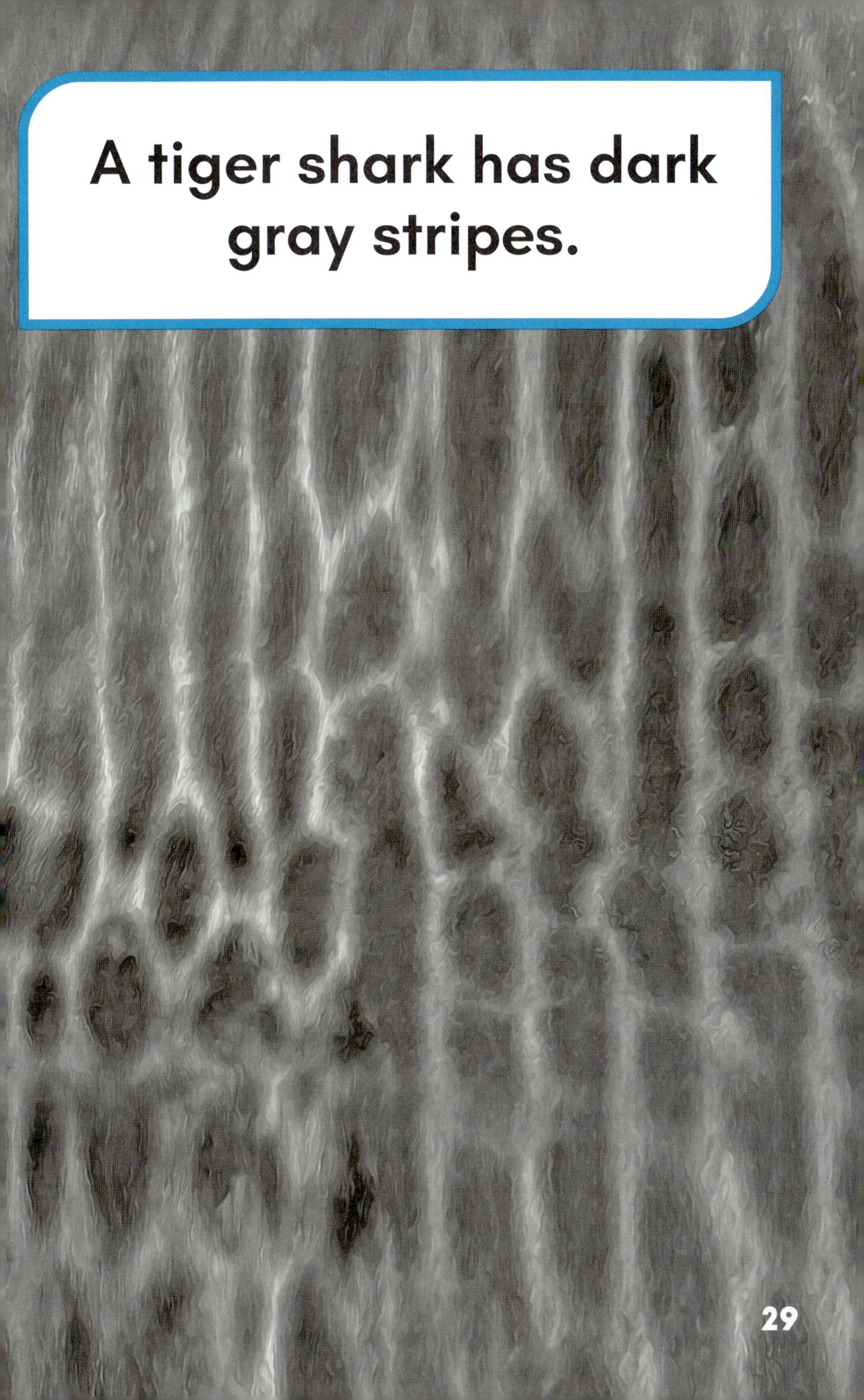

A tiger will never meet a tiger shark.

A tiger shark will never meet a tiger.

So which would you rather be? A tiger? Or a tiger shark?